I0490593

Thesis on Mercury in Motion

Ashai J D Parsons

Copyright © 2023 Ashai J D Parsons

All rights reserved.

ISBN: 9798391929512

DEDICATION

To my Mother and Father, who made sure to nurture any
interest I had while growing up

CONTENTS

Acknowledgments   i

Introduction   1

Plasma Converter
and
Tesla Turbine   4

Low to High
Pressure Gas Siphon   8

Main Shell and
Liquid
Compartment   11

Component Synergy   13

Electromagnetic
Field   15

Thermal Expansion   17

Centripetal Force and
Gyroscopes   20

Gyroscopic
Precession   23

Clausius-Clapeyron
Relation   28

Boyle's Law and   32
Gay-Lussac's

Charles's Law   35

Maxwell's Equations and 42
Particles in a Field

Momentum   47

The Lorentz Force   50

Bernoulli's Effect   53

Venturi Effect with Boiling 56
Points

Tying it Together   58

Hypothesis   64

Risk Assessment   66

Concepts Elsewhere   69

Concepts Continued   72

ACKNOWLEDGMENTS

Mr.Thompson
Ms.Taylor
Two teachers from Woolmer Hill who fed my passions

INTRODUCTION

Introduction to Low to High Pressure Siphon Systems with Plasma Converters and Tesla Turbines

In this introductory chapter, we will explore the intricate components of an advanced fluidic system that combines a low to high pressure siphon, a main shell and liquid compartment, and a plasma converter with a Tesla turbine. This innovative system harnesses the potential of fluid dynamics, thermodynamics, and electromagnetism to create an efficient and versatile energy conversion process. The following sections will outline the core principles and functionalities of each component.

1. Low to High Pressure Siphon

The low to high pressure siphon is a key component in this fluidic system, responsible for transferring fluid from a region of lower pressure to one of higher pressure. This is achieved through a combination of several curved pipes, and a sealed vacuum chamber. The pump creates a pressure differential, which forces the fluid to flow from low to high pressure regions. The vacuum chamber, in conjunction with the pump, helps maintain the pressure difference, allowing the siphon to operate efficiently.

2.   Main Shell and Liquid Compartment

The main shell and liquid compartment serve as the central hub of the fluidic system, housing the working fluid and connecting the various components. The main shell is designed to withstand high pressure and temperature fluctuations, ensuring the safety and durability of the system. The liquid compartment, on the other hand, is responsible for holding and managing the fluid used in the process. It features a temperature control system and is made of materials that can handle the chemical and thermal properties of the working fluid. This compartment is crucial for maintaining the fluid's temperature and viscosity within the optimal range, ensuring efficient energy conversion.

3.   Plasma Converter and Tesla Turbine

The plasma converter and Tesla turbine work in tandem to transform the energy contained in the working fluid into usable electrical power. The plasma converter first ionizes the fluid, creating a plasma state that carries a high level of kinetic and thermal energy. The ionized fluid is then directed into the Tesla turbine, a bladeless and highly efficient turbine that converts the energy of the plasma into mechanical energy. As the plasma flows through the turbine's smooth, closely spaced disks, the mechanical energy is transferred to the turbine's shaft, which then drives an electrical generator. The resulting electrical power can be utilized for various applications or stored for future use.

In conclusion, the advanced fluidic system we have outlined combines the principles of mechanical physics, fluid dynamics, thermodynamics, and electromagnetism to create an innovative and efficient energy conversion process. Understanding the functionalities of the low to high pressure siphon, main shell and liquid compartment, and plasma converter with a Tesla turbine is essential for further development and optimization of this technology.

PLASMA CONVERTER
AND
TESLA TURBINE

Plasma Converter and Tesla Turbine

Nikola Tesla, a brilliant inventor, discovered that his unique turbine design could achieve high speeds and efficiency when high-pressure vapor was passed through it. In our system, we use mercury vapor to power the Tesla Turbine. Mercury is an interesting choice because, besides driving the turbine at high speeds, it can also conduct electricity, allowing for the creation of plasma within the system.

For those unfamiliar with plasma, it is a state of matter where particles are charged, meaning they have either a positive or negative electric charge. In this system, the plasma flows at high speeds, which causes the charged particles to naturally separate based on their densities. Plasma is less dense than gases or liquids, so it is forced toward the centre of the spinning turbine.

When the plasma reaches the centre, it loses speed and momentum, which allows it to cool down and return to its liquid mercury state. However, the important part is that it still retains its electrical charge. This unique interaction between the plasma, the Tesla turbine, and the mercury vapor creates an innovative and efficient way to generate electricity that can be easily understood and appreciated by people from all walks of life.

One potential design for the construction of the Tesla turbine and plasma converter system involves the use of iron with copper plating as the primary material for the turbine's disks and a high tensile metal for the central axle. The choice of materials is crucial for the system's overall performance, durability, and efficiency.

The turbine disks, made from iron and coated with a layer of copper, offer several advantages. Iron is a strong and durable material that can handle the high speeds and pressures generated within the system. The copper plating on the iron disks serves two primary purposes: it enhances the electrical conductivity of the disks, which is vital for the creation and flow of plasma, and it reduces friction between the disks and the surrounding mercury vapor, increasing the overall efficiency of the turbine.

The central axle, which transfers the mechanical energy from the spinning turbine to an electrical generator, must be made from a high tensile metal to withstand the forces generated by the high-speed rotation. One possible choice for this material is Inconel, a family of nickel-chromium-based superalloys known for their excellent mechanical strength and resistance to oxidation and corrosion at high temperatures. Inconel alloys are commonly used in high-stress applications, such as aerospace and power generation, making them a suitable choice for the central axle in this system.

To construct this component, the following steps could be taken:

1. Fabrication of iron disks: The iron disks can be fabricated using techniques such as casting, forging, or machining to achieve the desired shape and dimensions.
2. Copper plating: After the iron disks are fabricated, they can be electroplated with a thin layer of copper. This process involves submerging the iron disks in a copper sulphate solution and applying an electric current, causing the copper ions in the solution to bond with the iron surface.
3. Assembly of the Tesla turbine: The copper-plated iron disks are then carefully spaced and mounted onto the high tensile central axle. The spacing between the disks must be precise to ensure optimal performance and efficiency.
4. Integration of the plasma converter: The plasma converter is incorporated into the system, connecting it with the Tesla turbine and the main shell and liquid compartment. This component is responsible for ionizing the mercury vapor, creating the plasma necessary for the energy conversion process.
5. Testing and fine-tuning: After the assembly and integration are complete, the system is tested to ensure its proper functioning, safety, and efficiency. Adjustments can be made to the design or materials as necessary to optimize performance.

By carefully selecting materials and following a meticulous fabrication process, the Tesla turbine and plasma converter component can be built to provide a highly efficient and durable solution for energy conversion.

LOW TO HIGH PRESSURE
GAS SIPHON

Low to High Pressure Gas Siphon

In this part of the engine, there's a unique component located in the middle of the main shell. Imagine several musical instruments, like French horns, connected together to form a doughnut or toroid shape around the central axle. This arrangement helps in controlling the flow of mercury within the system.

The tubes that make up this toroid have a larger opening at the input and a smaller opening at the output. This design helps increase the speed of the mercury as it flows through the tubes and fights against the spinning force created by the rotation of the turbine.

To make the system even more efficient, we take advantage of mercury's unique properties, such as its expansion coefficient and boiling point. When the mercury is subjected to low pressure, it expands and boils at a lower temperature. This change causes the mercury to transform from a liquid to a gas as it moves through the tubes at high speeds.

This gaseous mercury then enters the central plasma converter and Tesla turbine, where it is ionized and used to generate electricity. This simple explanation of the low to high pressure gas siphon provides a clear understanding of how it contributes to the overall function and efficiency of the engine.

The design of this engine takes advantage of the way objects behave when they are spinning, using principles of gyroscopic motion. As the mercury rotates within the engine, its mass is naturally pushed or attracted towards the input of the pressure siphon. This allows the mercury to flow efficiently through the system, transferring energy as it moves from the outer edges of the main shell to the central plasma converter.

Mercury has a unique property that makes it particularly useful in this engine. It can expand significantly before reaching its boiling point – up to 4.14 times its original volume. So, if you start with 10 cubic centimetres of mercury, it can expand to 41.1 cubic centimetres. This expansion helps facilitate the flow of mercury throughout the engine's cycle.

Additionally, the expansion of mercury reduces the speed needed within the pressure siphon to create a low enough pressure for mercury to become a vapor. This is important because when mercury turns into vapor, it can then be converted into plasma to complete the engine's cycle.

In simpler terms, the engine's design and the properties of mercury work together to create an efficient system that moves the mercury from the outer part of the engine to the centre, where it's converted into plasma and used to generate electricity.

MAIN SHELL AND LIQUID COMPARTMENT

Main Shell and Liquid Compartment

The main shell and liquid compartment are crucial parts of this engine, designed to hold and manage the mercury. The engine is shaped like a parabola, which is ideal for a gyroscopic system. This shape helps direct the mercury into the next compartment when it expands due to its unique properties.

When the engine is not in use, or when the mercury is in its liquid state, it rests within the main shell. During operation, the engine has two separate circuits: one where the direct current (DC) flows within the mercury chamber, and another where the engine reaches full heat and speed, activating the siphon in the closed system.

The centre of the compartment contains a pillar that holds the electromagnet, plasma chamber, and return valve.

Meanwhile, the liquid-to-gas converter is situated on the central plane of the compartment. The main shell is designed to be symmetrical, ensuring that the mercury follows a specific path related to vortex physics when it reaches high speeds.

The shell's material is carefully chosen so that it doesn't mix with mercury and has enough strength to withstand the forces within the engine. It consists of multiple layers to ensure stability, and the outer layer remains stationary even when the mercury is in motion. All components are thoroughly tested to ensure they can withstand the high temperatures and pressures encountered during operation.

In summary, the main shell and liquid compartment play essential roles in managing the mercury, providing a safe and efficient environment for the engine to function. The unique parabolic design and carefully selected materials ensure the engine's performance and longevity.

COMPONENT SYNERGY

Component Synergy

The unique disc-shaped engine is designed to serve two main purposes: generate gyroscopic force from the movement of mercury and create a powerful electromagnetic field using plasma at high speeds. This engine is made up of three major parts, along with a few minor components, that work together to achieve its goals.

1. Main Chamber: This part of the engine holds the mercury and maintains its rotation, generating the necessary gyroscopic force for stability.
2. Pressure Siphon: This component allows the mercury to change from a liquid to a gas, taking advantage of its unique properties and the system's design.
3. Plasma Converter: This vital part of the engine converts the gaseous mercury into plasma, which

generates a strong electromagnetic field when rotating at high speeds.

Additionally, there's a small electromagnet that helps initiate the movement of mercury when a direct current (DC) is applied. The overall goal of this engine is to create a powerful magnetic field while maintaining exceptional gyroscopic stability. This combination allows objects to "levitate" above the engine when it is operating.

The engine works by passing an electric current through hot metallic gas (plasma) and maintaining a stable magnetic field when spinning at high speeds. As the technology continues to develop, it is expected that there will be numerous improvements and applications for this innovative   engine.

ELECTROMAGNETIC FIELD

Electromagnetic field

An electromagnetic field (EMF) is a fundamental concept in physics that arises from the interaction between electric and magnetic forces. It is a combined field that propagates through space as a wave, carrying both electric and magnetic energy. EMFs play a crucial role in various physical phenomena and are the foundation for understanding electromagnetic radiation, including visible light, radio waves, microwaves, and X-rays.

To understand electromagnetic fields, one must first grasp the concepts of electric and magnetic fields separately. Electric fields are generated by electric charges (either stationary or in motion) and exert forces on other charges within their vicinity. The strength and direction of the electric field are determined by the arrangement and magnitude of these charges.

Magnetic fields, on the other hand, are produced by moving electric charges (i.e., electric currents) and act on other moving charges within their influence. They can also be generated by changing electric fields. Magnetic fields have both a magnitude and a direction, much like electric fields.

The relationship between electric and magnetic fields is described by Maxwell's equations, a set of four fundamental equations that govern electromagnetism. These equations reveal that time-varying electric fields can produce magnetic fields and vice versa. This interdependence leads to the generation of electromagnetic fields, which propagate as waves through space.

Electromagnetic waves consist of oscillating electric and magnetic fields that are perpendicular to each other and the direction of propagation. They travel at the speed of light (approximately 299,792 kilometres per second in a vacuum) and can transport energy from one point to another. The properties of these waves, such as frequency and wavelength, define the different types of electromagnetic radiation within the electromagnetic spectrum.

In conclusion, an electromagnetic field is a combined field that results from the interaction of electric and magnetic fields. It is a cornerstone concept in physics, playing a vital role in various physical phenomena and technologies, such as wireless communication, medical imaging, and power generation. A deep understanding of electromagnetic fields is essential for college graduates pursuing careers in fields like engineering, physics, and telecommunications.

THERMAL EXPANSION

Thermal Expansion

Thermal expansion is the phenomenon in which a material's volume changes as a function of temperature. This occurs because the constituent particles of a material, such as atoms or molecules, experience increased kinetic energy when the temperature rises. As a result, they vibrate more vigorously and move further apart from each other, causing the material to expand. Conversely, when the temperature drops, the kinetic energy of the particles decreases, and they move closer together, leading to contraction. The extent of thermal expansion or contraction is determined by a material's coefficient of thermal expansion, which is a property specific to each substance.

Mercury (Hg), a heavy metal and the only liquid metal at room temperature, exhibits unique thermal expansion properties that are particularly relevant in scientific and engineering applications. Mercury has a high coefficient of linear thermal expansion, which means it expands significantly in response to temperature changes. This property is utilized in various temperature-measuring instruments, such as mercury-in-glass thermometers, where the expansion of mercury in a glass column is directly proportional to the change in temperature.

Mercury's thermal expansion can be described by the following equation:

$$\Delta V = V_0 * \beta * \Delta T$$

where:

- ΔV is the change in volume,
- V_0 is the initial volume,
- β is the coefficient of volume expansion for mercury (approximately 1.8×10^{-4} K^{-1}), and
- ΔT is the change in temperature.

The high coefficient of volume expansion for mercury implies that even small temperature changes can result in significant volume changes. This property makes mercury an excellent candidate for use in devices that require precise temperature measurements, such as laboratory equipment and specialized thermometers.

However, mercury's thermal expansion properties also pose challenges in certain applications. For example, when mercury is used in electrical switches or other temperature-sensitive equipment, the significant expansion or contraction of mercury with temperature fluctuations can lead to operational issues or even failure of the equipment. Therefore, engineers and scientists must carefully consider the implications of mercury's thermal expansion properties when designing systems that involve mercury.

In summary, thermal expansion is a fundamental physical phenomenon that results from changes in the temperature of a material. Mercury, due to its high coefficient of thermal expansion, exhibits significant volume changes in response to temperature fluctuations. This property has both advantages and drawbacks, which must be carefully considered in the design and operation of scientific and engineering systems that utilize mercury.

Mercury's volume coefficient of expansion is 0.00018, so it expands by 0.018 percent in volume for every degree of temperature increase. Mercury has a boiling point of 356.7 Degrees C. Assuming Room temperature is 26.7 Degrees, mercury can expand 4.14 times before boiling

CENTRIPETAL FORCE
AND
GYROSCOPES

Centripetal Force and Gyroscopes

Centripetal force is a fundamental concept in classical mechanics, referring to the force required to keep an object moving in a circular path. The word "centripetal" is derived from the Latin words "centrum" (centre) and "petere" (to seek), indicating that centripetal force always acts toward the centre of the circular path. This force is essential for understanding the behaviour of various rotating systems, including gyroscopes.

A gyroscope is a spinning wheel or disk mounted on a set of gimbals, allowing it to rotate freely in three dimensions. The primary characteristic of a gyroscope is its ability to maintain its rotational axis orientation despite external forces or torques, a property known as gyroscopic stability or rigidity in space. This stability arises from the conservation of angular momentum, which is a conserved quantity in the absence of external torques.

The centripetal force in a gyroscope can be analysed by considering a small mass element located at a distance 'r' from the axis of rotation. As the gyroscope spins, this mass element moves in a circular path with a constant angular velocity 'ω'. According to Newton's second law of motion, the centripetal force acting on this mass element is given by:

$F\_c = m * a\_c$

where:

- F_c is the centripetal force,
- m is the mass of the element, and
- a_c is the centripetal acceleration.

The centripetal acceleration can be expressed in terms of angular velocity as:

$a\_c = r * \omega^2$

Substituting this into the centripetal force equation, we get:

$F\_c = m * r * \omega^2$

This equation shows that the centripetal force acting on the mass element is directly proportional to its distance from the axis of rotation and the square of the angular velocity. It is important to note that the centripetal force is always directed toward the centre of rotation, keeping the mass element on its circular path.

In the context of a gyroscope, the centripetal force acts on each mass element within the spinning wheel, contributing to the gyroscopic stability. The conservation of angular momentum and the centripetal forces acting on the individual mass elements work together to resist external torques and maintain the gyroscope's rigidity in space.

In summary, centripetal force is a key concept in understanding the behaviour of gyroscopes, as it helps maintain the circular motion of individual mass elements in the spinning wheel. The gyroscopic stability arises from the interplay between centripetal forces and the conservation of angular momentum, which allows gyroscopes to maintain their orientation despite external forces. This property has numerous applications in engineering and science, including navigation systems, attitude control in spacecraft, and stabilization of cameras and vehicles.

GYROSCOPIC PRECESSION

Gyroscopic Precession

Gyroscopic precession is a crucial concept in classical mechanics that describes the behaviour of a rotating body, such as a gyroscope or spinning top, when subjected to an external torque. This phenomenon is essential for understanding the dynamics of various systems, including aircraft and spacecraft attitude control, bicycle stability, and the Earth's precession.

To comprehend gyroscopic precession, we must first understand that a gyroscope is a spinning wheel or disk mounted on a set of gimbals, enabling it to rotate freely in three dimensions. The main property of a gyroscope is its ability to maintain its rotational axis orientation despite external forces or torques, known as gyroscopic stability or rigidity in space. This stability arises from the conservation of angular momentum.

When an external torque is applied to a spinning gyroscope, the gyroscope's response is counterintuitive: instead of rotating in the direction of the applied torque, it rotates in a direction perpendicular to both the torque and its initial angular momentum vector. This rotation is called gyroscopic precession.

The rate of precession (Ω_p) is given by the following equation:

$$\Omega\_p = T / (L * \sin(\theta))$$

where:

- Ω_p is the precession rate,
- T is the magnitude of the applied torque,
- L is the magnitude of the gyroscope's angular momentum, and
- θ is the angle between the torque and angular momentum vectors.

Gyroscopic precession is a direct consequence of the conservation of angular momentum. When an external torque is applied, the gyroscope's angular momentum changes in a direction perpendicular to both the torque and the initial angular momentum vector. This results in the gyroscope's rotational axis changing its orientation, causing precession.

The phenomenon of gyroscopic precession is observed in various real-world applications:

1. In aviation, gyroscopic instruments such as attitude indicators and heading indicators exploit

gyroscopic precession to maintain an aircraft's orientation information.

2. The Earth experiences gyroscopic precession due to the gravitational interaction with the Sun and Moon. This causes the Earth's rotational axis to precess over a period of approximately 26,000 years, affecting the positions of celestial objects in the sky over time.

3. Gyroscopic precession plays a significant role in the stability of bicycles and motorcycles. The spinning wheels act as gyroscopes, and the precessional forces generated when the rider leans into a turn help maintain balance and control.

In summary, gyroscopic precession is a key concept in understanding the behaviour of rotating systems when subjected to external torques. It is a direct consequence of the conservation of angular momentum and causes the rotation axis to change its orientation in a direction perpendicular to both the applied torque and the initial angular momentum vector. Gyroscopic precession has numerous applications in engineering and science, from aircraft instruments to the stability of bicycles and the Earth's precession.

Torque-free precession, also known as free precession or inertial precession, is a specific type of gyroscopic precession that occurs when a spinning body is not subjected to any external torques. This phenomenon is essential for understanding the behaviour of various rotating systems, including spacecraft attitude control, the motion of planets, and the dynamics of spinning celestial bodies like pulsars and black holes.

Torque-free precession is a consequence of the conservation of angular momentum in the absence of external torques. In this case, the spinning body's angular momentum is conserved, but the orientation of the angular momentum vector may change due to the object's asymmetric mass distribution or the presence of internal forces.

For a spinning body with principal moments of inertia I_1, I_2, and I_3, and angular velocities ω_1, ω_2, and ω_3 about its principal axes, the conservation of angular momentum can be expressed as:

$$I_1 * \omega_1 = \text{constant}_1$$
$$I_2 * \omega_2 = \text{constant}_2$$
$$I_3 * \omega_3 = \text{constant}_3$$

When the moments of inertia are not equal ($I_1 \neq I_2 \neq I_3$), the angular velocities ω_1, ω_2, and ω_3 will change over time, causing the rotation axis to precess. The rate of torque-free precession can be calculated using the Euler equations of motion for a rigid body:

$$(I_2 - I_3) * \omega_2 * \omega_3 / I_1 = d\omega_1/dt$$
$$(I_3 - I_1) * \omega_3 * \omega_1 / I_2 = d\omega_2/dt$$
$$(I_1 - I_2) * \omega_1 * \omega_2 / I_3 = d\omega_3/dt$$

These equations describe how the angular velocities change over time, leading to torque-free precession.

One notable example of torque-free precession is the motion of a spacecraft in orbit. In the absence of external torques, the spacecraft's angular momentum is conserved, but its orientation may change due to the varying angular velocities about its principal axes. This effect is crucial for attitude control in spacecraft, and mission planners must account for torque-free precession when designing control systems.

In summary, torque-free precession is a type of gyroscopic precession that occurs in the absence of external torques. It arises from the conservation of angular momentum and is governed by the Euler equations of motion for a rigid body. Torque-free precession is crucial for understanding the behaviour of various rotating systems, such as spacecraft attitude control and the motion of celestial bodies, and has numerous applications in engineering and science.

This is the type of Gyroscopic Precession the mercury will be   subject   to.

CLAUSIUS-CLAPEYRON RELATION

The Clausius-Clapeyron relation

The Clausius-Clapeyron relation is an essential concept in thermodynamics and physical chemistry, providing a mathematical framework to describe the relationship between boiling points and vapor pressures of a substance undergoing a phase transition between two phases. It is particularly useful for understanding the phase behaviour of pure substances, such as the equilibrium between liquid and vapor phases, and determining the temperature or pressure dependence of phase transitions.

The Clausius-Clapeyron relation is derived from the principles of thermodynamics, specifically, by considering the changes in enthalpy, entropy, and Gibbs free energy during a phase transition. It is expressed as:

$$dP/dT = L / (T * \Delta v)$$

where:

- dP/dT is the rate of change of vapor pressure (P) with respect to temperature (T),
- L is the latent heat of the phase transition (positive for vaporization, negative for condensation),
- T is the absolute temperature, and
- Δv is the change in specific volume between the two phases (usually the difference between the specific volume of the vapor and the specific volume of the liquid).

In many practical situations, the change in specific volume (Δv) is much larger for the vapor phase than for the liquid phase. This allows us to simplify the Clausius-Clapeyron relation using the ideal gas law for the vapor phase:

$$PV = nRT$$

where:

- P is the vapor pressure,
- V is the molar volume of the vapor,
- n is the number of moles,
- R is the ideal gas constant, and
- T is the absolute temperature.

By assuming the liquid phase to be incompressible (constant specific volume) and the vapor phase to obey the ideal gas law, the Clausius-Clapeyron relation can be further simplified to:

$$d(\ln P) / d(1/T) = -L / R$$

This equation is particularly useful in determining the

temperature dependence of vapor pressure, which is essential for understanding boiling points, distillation, and the behaviour of substances in various industrial and laboratory processes.

To summarize, the Clausius-Clapeyron relation is a fundamental concept in thermodynamics that provides a mathematical framework for understanding the relationship between boiling points and vapor pressures of a substance undergoing phase transitions. Derived from the principles of thermodynamics, this relation is particularly useful for studying the equilibrium between liquid and vapor phases and determining the temperature or pressure dependence of phase transitions. The Clausius-Clapeyron relation is widely employed in various scientific and engineering applications, including distillation, material characterization, and process design.

The Clausius-Clapeyron relation, as previously described, is crucial for understanding the relationship between boiling points and vapor pressures during phase transitions. When considering liquid mercury turning into vapor with an increase in velocity, we can still apply the same principles from the Clausius-Clapeyron relation. However, it is essential to account for the fact that an increase in velocity impacts the pressure and, consequently, the boiling point of the substance.

In the case of liquid mercury, as the velocity increases, it experiences a decrease in pressure due to the Bernoulli's principle, which states that an increase in the speed of a fluid occurs simultaneously with a decrease in pressure. When the pressure is reduced, the boiling point of the mercury also decreases, causing the liquid mercury to vaporize at lower temperatures than under standard atmospheric conditions. This behaviour can be analysed using the Clausius-Clapeyron relation.

The simplified Clausius-Clapeyron equation for mercury vaporization can be expressed as:

$$d(\ln P) / d(1/T) = -L / R$$

where:

- P is the vapor pressure of mercury,
- T is the absolute temperature,
- L is the latent heat of vaporization for mercury, and
- R is the ideal gas constant.

By solving this equation, we can determine the relationship between temperature and vapor pressure for mercury. Additionally, we can estimate the boiling point of mercury at various pressures, enabling us to analyse the phase transition of liquid mercury to vapor as the velocity increases and pressure decreases.

In conclusion, the Clausius-Clapeyron relation is vital for understanding the phase behaviour of liquid mercury turning into vapor with an increase in velocity. The decrease in pressure caused by the increased velocity, according to Bernoulli's principle, results in a lowered boiling point for mercury, which can be analysed using the Clausius-Clapeyron relation. This relationship enables us to determine the temperature and pressure conditions under which mercury transitions from the liquid to the vapor phase, providing valuable insights into the behaviour of mercury in various industrial and scientific applications.

BOYLE'S LAW AND GAY-LUSSAC

Boyle's Law and Gay-Lussac

Boyle's Law and Gay-Lussac's Law are fundamental gas laws in the field of thermodynamics and physical chemistry, describing the behaviour of ideal gases under varying conditions of pressure, volume, and temperature. Both laws are derived from the ideal gas law, which is a simplified model to describe the behaviour of real gases at relatively low pressures and high temperatures.

Boyle's Law: Boyle's Law describes the relationship between the pressure and volume of a given amount of gas at constant temperature. It is named after Robert Boyle, who first formulated this law in 1662. Mathematically, Boyle's Law is expressed as:

$$P1 * V1 = P2 * V2$$

where:

- P1 and P2 represent the initial and final pressures of the gas, respectively,
- V1 and V2 represent the initial and final volumes of the gas, respectively, and
- Temperature (T) is held constant.

Boyle's Law states that the pressure of an ideal gas is inversely proportional to its volume when the temperature and the amount of gas are kept constant. This means that as the volume of a gas decreases, the pressure increases, and vice versa. This principle is crucial for understanding various phenomena in chemistry, physics, and engineering, such as gas compression, piston movement in engines, and respiration in living organisms.

Gay-Lussac's Law: Gay-Lussac's Law, also known as the pressure-temperature law or Amontons' Law, describes the relationship between the pressure and temperature of a given amount of gas at constant volume. It is named after French chemist Joseph Louis Gay-Lussac, who first formulated this law in 1802. Mathematically, Gay-Lussac's Law is expressed as:

$(P1 / T1) = (P2 / T2)$

where:

- P1 and P2 represent the initial and final pressures of the gas, respectively,
- T1 and T2 represent the initial and final absolute temperatures of the gas, respectively, and
- Volume (V) is held constant.

Gay-Lussac's Law states that the pressure of an ideal gas is directly proportional to its absolute temperature when the volume and the amount of gas are kept constant. This means that as the temperature of a gas increases, its pressure also increases, and vice versa. This principle is essential for understanding various processes involving heating or cooling gases, such as tire pressure changes with temperature, gas storage and transportation, and the behaviour of gases in chemical reactions.

In summary, Boyle's Law and Gay-Lussac's Law are fundamental gas laws that describe the behaviour of ideal gases under varying conditions of pressure, volume, and temperature. Boyle's Law focuses on the inverse relationship between pressure and volume at constant temperature, while Gay-Lussac's Law examines the direct relationship between pressure and temperature at constant volume. Both laws are vital for understanding various phenomena in chemistry, physics, and engineering, and are derived from the ideal gas law, which is a simplified model used to describe the behaviour of real gases at relatively low pressures and high temperatures.

CHARLES'S LAW

Charles's Law

Charles's Law, also known as the volume-temperature law, is a fundamental gas law that describes the behaviour of ideal gases under varying conditions of volume and temperature while keeping pressure constant. It is named after Jacques Charles, a French chemist, and physicist who formulated this law in the 1780s. Charles's Law is derived from the ideal gas law, which is a simplified model used to describe the behaviour of real gases at relatively low pressures and high temperatures.

Charles's Law states that the volume of an ideal gas is directly proportional to its absolute temperature when the pressure and the amount of gas are kept constant. In other words, as the temperature of a gas increases, its volume also increases, and vice versa, provided that the pressure remains unchanged. Mathematically, Charles's Law can be expressed as:

$$(V1 / T1) = (V2 / T2)$$

where:

- V1 and V2 represent the initial and final volumes of the gas, respectively,
- T1 and T2 represent the initial and final absolute temperatures of the gas, respectively, and
- Pressure (P) is held constant.

It is important to note that the temperatures in Charles's Law must be expressed in Kelvin (K), the absolute temperature scale, to maintain the proportionality between volume and temperature.

Charles's Law is crucial for understanding various phenomena in chemistry, physics, and engineering, such as the behaviour of gases in balloons, the expansion and contraction of materials due to temperature changes, and the design of thermal systems like heat exchangers and refrigeration cycles.

In summary, Charles's Law is a fundamental gas law that describes the relationship between the volume and temperature of an ideal gas when pressure is kept constant. It is derived from the ideal gas law, which is a simplified model used to describe the behaviour of real gases at relatively low pressures and high temperatures. Charles's Law is essential for understanding various phenomena in chemistry, physics, and engineering, as it provides insight into the behaviour of gases under varying conditions of volume and temperature while maintaining constant pressure.

ELECTRICAL RESISITVITY
AND
CONDUCTIVITY

Electrical Resistivity and Conductivity

Electrical resistivity and conductivity are essential properties
of materials that describe their ability to oppose or conduct
the flow of electric current, respectively. These properties
are particularly important when considering the behaviour
of metals like mercury in various applications, such as
electrical circuits, sensors, and switches.

Electrical Resistivity: Electrical resistivity (ϱ) is a material
property that quantifies the opposition a material offers to
the flow of electric current. It is the inverse of electrical
conductivity and is typically measured in ohm-meters
($\Omega \cdot m$). The resistivity of a material depends on its atomic
structure, temperature, and impurities or defects present.
For metals, including mercury, resistivity generally increases
with increasing temperature due to increased lattice
vibrations, which cause more frequent collisions between
the charge carriers (electrons) and the lattice ions.

The resistivity of a material can be calculated using the formula:

$$\varrho = R * (A / L)$$

where:

- ϱ is the resistivity,
- R is the electrical resistance,
- A is the cross-sectional area of the material, and
- L is the length of the material.

Electrical Conductivity: Electrical conductivity (σ) is a material property that quantifies the ability of a material to conduct electric current. It is the inverse of electrical resistivity and is typically measured in Siemens per meter (S/m). The conductivity of a material depends on the density and mobility of its charge carriers, which, for metals like mercury, are the valence electrons.

The conductivity of a material can be calculated using the formula:

$$\sigma = 1 / \varrho$$

where:

- σ is the electrical conductivity, and
- ϱ is the electrical resistivity.

Focusing on Mercury: Mercury is a unique metal due to its liquid state at room temperature. It has relatively high electrical conductivity compared to other non-alkali metals, making it useful in certain electrical applications. However, its high electrical conductivity also means that it has a relatively low electrical resistivity. The electrical resistivity of mercury at 20°C is approximately 9.8×10^{-8} Ω·m.

Mercury's electrical properties can change significantly with temperature. As the temperature increases, the resistivity of mercury also increases due to the increased lattice vibrations, as mentioned earlier. This behaviour is essential to consider in applications where mercury is exposed to a wide range of temperatures or used as a temperature-sensitive element.

In summary, electrical resistivity and conductivity are critical properties of materials that describe their ability to oppose or conduct the flow of electric current. Mercury, as a unique liquid metal, exhibits relatively high electrical conductivity and low resistivity, which are both temperature-dependent. These properties are essential for understanding the behaviour of mercury in various applications, such as electrical circuits, sensors, and switches, as well as for designing systems that take advantage of its unique electrical characteristics.

At elevated temperatures, the electrical properties of mercury change, and understanding these changes is crucial for applications involving mercury vapor or heated mercury.

Mercury at 356.7°C: At 356.7°C (674°F), mercury is in its liquid state and close to its boiling point (356.73°C, or 674.1°F). As the temperature increases, the resistivity of mercury increases due to the enhanced lattice vibrations that cause more frequent collisions between the charge carriers (electrons) and the lattice ions. The precise resistivity value at 356.7°C is not readily available; however, it will be higher than the resistivity at room temperature (approximately 9.8×10^{-8} Ω·m at 20°C). This increased resistivity results in a decrease in electrical conductivity, making heated mercury less effective as an electrical conductor than at room temperature.

Mercury Vapor: When mercury reaches its boiling point, it transitions from a liquid to a gaseous state, becoming mercury vapor. The electrical properties of mercury vapor are significantly different from those of liquid mercury. In the vapor phase, mercury atoms are ionized, creating a mixture of electrons and positively charged mercury ions. This ionized gas, or plasma, exhibits electrical conductivity as the electrons and ions can move freely and carry electric charge.

The electrical resistivity of mercury vapor depends on several factors, such as temperature, pressure, and ionization degree. At higher temperatures and pressures, the degree of ionization increases, leading to higher electrical conductivity and lower resistivity. However, the resistivity values for mercury vapor are not constant and can vary widely depending on the specific conditions.

In summary, the electrical properties of mercury change significantly at elevated temperatures and in its vapor phase. At 356.7°C, liquid mercury exhibits increased resistivity and decreased conductivity compared to room temperature. In the vapor phase, ionized mercury atoms create a plasma with electrical conductivity that depends on temperature, pressure, and ionization degree. Understanding these changes is essential for applications involving heated mercury or mercury vapor, such as high-intensity discharge lamps, mercury vapor rectifiers, and thermoelectric devices.

MAXWELL'S EQUATIONS AND PARTICLES IN A FIELD

Maxwell's Equations and Particles in a Field

Maxwell's equations are a set of four fundamental equations that describe the behaviour of electric and magnetic fields and their interactions with matter. These equations, formulated by James Clerk Maxwell in the 19th century, laid the foundation for classical electromagnetism and led to the development of the theory of electromagnetic waves. Maxwell's equations are essential for understanding various phenomena in physics, engineering, and telecommunications, such as radio waves, microwaves, and light.

The four Maxwell's equations are:

1. Gauss's Law for Electric Fields: This law states that the electric flux through a closed surface is

proportional to the net electric charge enclosed by the surface. Mathematically, it is expressed as:

$$\oint E \cdot dA = Q\_enclosed / \varepsilon_0$$

where:

- E is the electric field,
- dA is an infinitesimal area vector on the closed surface,
- Q_enclosed is the total charge enclosed by the surface, and
- ε_0 is the vacuum permittivity.

2. Gauss's Law for Magnetic Fields: This law states that the magnetic flux through a closed surface is always zero, which implies that there are no magnetic monopoles and magnetic field lines are always closed loops. Mathematically, it is expressed as:

$$\oint B \cdot dA = 0$$

where:

- B is the magnetic field,
- dA is an infinitesimal area vector on the closed surface.

3. Faraday's Law of Electromagnetic Induction: This law relates the change in the magnetic field with time to the electromotive force (EMF) induced in a closed loop. Mathematically, it is expressed as:

$$\oint E \cdot dl = - \, d(\int B \cdot dA) \, / \, dt$$

where:

- E is the electric field,
- dl is an infinitesimal vector along the closed loop,
- B is the magnetic field,
- dA is an infinitesimal area vector within the loop, and
- t is time.

4. Ampère's Law with Maxwell's Addition (Maxwell-Ampère Law): This law relates the electric current and the time-varying electric field to the magnetic field around a closed loop. Mathematically, it is expressed as:

$$\oint B \cdot dl = \mu_0 \, (I\_enclosed + \varepsilon_0 \, d(\int E \cdot dA) \, / \, dt)$$

where:

- B is the magnetic field,
- dl is an infinitesimal vector along the closed loop,
- μ_0 is the vacuum permeability,
- I_enclosed is the total current enclosed by the loop,
- E is the electric field,
- dA is an infinitesimal area vector within the loop, and
- t is time.

Particles in a Field: Maxwell's equations can also be applied to understand the behaviour of particles in electric and magnetic fields. Charged particles, such as electrons and protons, interact with these fields, leading to various phenomena like the Lorentz force, which causes charged particles to experience a force in the presence of electric and magnetic fields. The motion of charged particles in fields can be described using the Lorentz force equation:

$$F = q(E + v \times B)$$

where:

- F is the Lorentz force experienced by the particle,
- q is the charge of the particle,
- E is the electric field,
- v is the velocity of the particle, and
- B is the magnetic field.

In summary, Maxwell's equations are a set of four fundamental equations that describe the behaviour of electric and magnetic fields and their interactions with matter. These equations are essential for understanding various phenomena in physics, engineering, and telecommunications, and they can also be applied to understand the behaviour of particles in electric and magnetic fields. The motion of charged particles in these fields can be described using the Lorentz force equation, which dictates the force experienced by charged particles in the presence of electric and magnetic fields.

Understanding the behaviour of particles in fields has led to significant advancements in various fields of physics and engineering. For example, particle accelerators like the Large Hadron Collider (LHC) use strong electric and magnetic fields to accelerate charged particles to nearly the speed of light. These high-energy particles can then be collided to probe the fundamental properties of matter and the forces that govern their interactions.

Similarly, in plasma physics, the study of charged particles in electric and magnetic fields is essential for understanding the behaviour of ionized gases, which have applications in fusion energy, space propulsion, and astrophysics.

In the context of electromagnetic radiation, Maxwell's equations predict the existence of electromagnetic waves that can propagate through space. These waves carry energy and momentum and can exert forces on charged particles. This has important implications for understanding the interaction of light with matter, leading to phenomena such as the photoelectric effect, Compton scattering, and pair production.

Additionally, Maxwell's equations play a crucial role in understanding antenna theory, where the interaction between charged particles and electromagnetic fields is essential for the transmission and reception of radio waves.

In conclusion, Maxwell's equations are fundamental to understanding electric and magnetic fields and their interactions with charged particles. These equations have far-reaching applications in various fields of physics, engineering, and telecommunications, from particle accelerators and plasma physics to antenna theory and the behaviour of light. By studying Maxwell's equations and the motion of charged particles in fields, researchers continue to make ground-breaking discoveries and develop new technologies that shape our understanding of the universe and improve our everyday lives.

MOMENTUM

Momentum

Momentum is a fundamental concept in Newtonian mechanics and is defined as the product of an object's mass and its velocity. It is a vector quantity, meaning it has both magnitude and direction. Mathematically, momentum (p) can be represented as:

$p = m * v$

where:

- p is the momentum,
- m is the mass of the object, and
- v is the object's velocity vector.

In classical mechanics, momentum is conserved in a closed system, meaning that the total momentum of the system remains constant if no external forces act upon it. This conservation law is derived from Newton's second and third laws of motion.

Newton's second law states that the force acting on an object is equal to the time rate of change of its momentum. Mathematically, it can be expressed as:

$F = dp/dt$

where:

- F is the net force acting on the object, and
- dp/dt is the time derivative of the momentum.

This equation shows that when a net force acts on an object, its momentum will change over time, and the object will experience acceleration. In cases where the mass of the object is constant, the equation can be simplified to the more familiar form:

$F = m * a$

where:

- a is the acceleration experienced by the object.

Newton's third law states that for every action, there is an equal and opposite reaction. This implies that when two objects interact, the forces they exert on each other are equal in magnitude but opposite in direction. In the context of momentum, this means that the total momentum of a closed system is conserved during interactions between objects, as the momentum gained by one object is equal to the momentum lost by the other.

The conservation of momentum is a powerful tool for analysing collisions and interactions between objects. In particular, it is useful for understanding elastic and inelastic collisions. In an elastic collision, both momentum and kinetic energy are conserved, while in an inelastic collision, only momentum is conserved, and some kinetic energy is converted to other forms of energy, such as heat or deformation.

In summary, momentum is a key concept in Newtonian mechanics and is defined as the product of an object's mass and velocity. The conservation of momentum is a fundamental principle derived from Newton's laws of motion and is crucial for understanding various phenomena in classical mechanics, such as collisions and interactions between objects. By analysing the conservation of momentum in different situations, researchers can gain valuable insights into the behaviour of physical systems and develop a deeper understanding of the fundamental principles governing the motion of objects.

THE LORENTZ FORCE

The Lorentz Force

The Lorentz force is a fundamental concept in electromagnetism that describes the force experienced by a charged particle moving through electric and magnetic fields. It plays a crucial role in understanding the behaviour of charged particles in various contexts, from the motion of electrons in electrical circuits to the trajectories of charged particles in particle accelerators and plasmas.

Mathematically, the Lorentz force (F) acting on a charged particle is given by:

$$F = q(E + v \times B)$$

where:

- F is the Lorentz force experienced by the particle,
- q is the charge of the particle,

- E is the electric field vector,
- v is the velocity vector of the particle, and
- B is the magnetic field vector.

The first term, qE, represents the force due to the electric field acting on the charged particle. This force is directly proportional to the charge of the particle and the strength of the electric field and acts along the direction of the electric field.

The second term, $q(v \times B)$, represents the force due to the magnetic field acting on the charged particle. This force is perpendicular to both the velocity vector of the particle and the magnetic field vector. The direction of the magnetic force can be determined using the right-hand rule.

The right-hand rule is a mnemonic tool used to find the direction of the magnetic force on a charged particle moving through a magnetic field. To apply the right-hand rule, follow these steps:

1. Point your right-hand thumb in the direction of the velocity vector of the charged particle (v).
2. Extend your fingers in the direction of the magnetic field vector (B).
3. The direction in which your palm pushes is the direction of the magnetic force on a positive charge. If the charge is negative, the force will be in the opposite direction.

It is important to note that the magnetic force acts perpendicular to the particle's velocity, causing it to change direction but not its speed. As a result, the magnetic force does no work on the charged particle, and its kinetic energy remains constant.

In summary, the Lorentz force is a fundamental concept in electromagnetism that describes the force experienced by a charged particle moving through electric and magnetic fields. The right-hand rule is a mnemonic tool used to determine the direction of the magnetic force acting on the particle. Understanding the Lorentz force and the right-hand rule is crucial for analysing the behaviour of charged particles in various physical systems, including electrical circuits, particle accelerators, and plasmas.

BERNOULLI'S PRINCIPLE

Bernoulli's Principle – Low to High Pressure Siphon

Bernoulli's principle is a fundamental concept in fluid dynamics that describes the relationship between the pressure, velocity, and elevation of a fluid in a steady-state, incompressible flow. It is derived from the conservation of energy and can be used to analyse various fluid flow phenomena, including the behaviour of low to high-pressure siphons.

According to Bernoulli's principle, the sum of the pressure energy, kinetic energy, and potential energy per unit volume of a fluid remains constant along a streamline, assuming negligible viscous effects and heat transfer. Mathematically, this can be expressed as:

$$P + 0.5\varrho v^2 + \varrho gh = \text{constant}$$

where:

- P is the pressure in the fluid,
- ϱ is the fluid density,
- v is the fluid velocity,
- g is the acceleration due to gravity, and
- h is the elevation of the fluid above a reference level.

Bernoulli's principle implies that as the fluid velocity increases, the pressure in the fluid decreases, and vice versa. This relationship is fundamental to understanding the behaviour of low to high-pressure siphons.

A siphon is a device used to transfer fluid from a higher elevation to a lower elevation without the need for pumping. In the context of a low to high-pressure siphon, the fluid flows from a region of low pressure to a region of high pressure. This might seem counterintuitive since fluids typically flow from high to low pressure; however, this behaviour can be explained using Bernoulli's principle.

In a low to high-pressure siphon, the fluid is initially accelerated due to a pressure difference or other driving force, such as gravity. As the fluid velocity increases, its pressure decreases according to Bernoulli's principle. Consequently, the fluid flows from the low-pressure region to the high-pressure region.

One example of a low to high-pressure siphon is the Venturi effect, in which a constriction in a fluid-carrying pipe causes the fluid velocity to increase and the pressure to decrease. This pressure drop can be used to draw fluid from a secondary source, such as a reservoir, into the primary flow.

In summary, Bernoulli's principle is a fundamental concept

in fluid dynamics that describes the relationship between the pressure, velocity, and elevation of a fluid in a steady-state, incompressible flow. This principle can be used to analyse the behaviour of low to high-pressure siphons, where fluid flows from a region of low pressure to a region of high pressure. Understanding Bernoulli's principle and its applications to siphon systems is essential for solving various fluid flow problems in engineering and physics.

VENTURI EFFECT WITH BOILING POINTS

Venturi effect with boiling points

The Venturi effect is a phenomenon in fluid dynamics that occurs when a fluid flows through a constricted section of a pipe, causing a decrease in pressure and an increase in velocity. This effect has a significant impact on the boiling points of fluids and can be used to facilitate phase changes in certain applications.

According to the Clausius-Clapeyron relation, the boiling point of a fluid is directly related to its pressure. When the pressure of a fluid decreases, its boiling point also decreases, and vice versa. Consequently, the Venturi effect can be used to induce boiling in a fluid by reducing its pressure.

In a system designed to exploit the Venturi effect and boiling points, fluid flows through a pipe with a constricted section, where the fluid velocity increases, and the pressure decreases. If the pressure reduction is significant enough, the fluid's boiling point may drop below its current temperature, causing it to vaporize.

For example, consider a system where water flows through a pipe with a constriction. As the water passes through the constriction, its pressure decreases due to the Venturi effect. If the pressure reduction is sufficient, the boiling point of the water may drop below its current temperature, causing it to boil and turn into steam.

This phenomenon can be useful in various industrial applications, such as steam generation, refrigeration, and vacuum distillation. By carefully designing the pipe geometry and controlling the fluid flow rate, engineers can manipulate the boiling points of fluids to achieve desired phase changes and optimize system performance.

In summary, the Venturi effect can significantly influence the boiling points of fluids by inducing pressure changes as the fluid flows through a constricted pipe section. By exploiting this effect, engineers can induce phase changes in fluids and optimize the performance of various systems, such as steam generators, refrigeration units, and vacuum distillation processes. Understanding the relationship between the Venturi effect and boiling points is essential for solving complex fluid flow problems and designing efficient systems in engineering and physics.

TYING IT TOGETHER

When mercury is set in motion by a direct current (DC) in the presence of an electromagnetic field, various phenomena come into play, including the Lorentz force, electrical resistivity, and mercury's expansion coefficient. The Lorentz force, acting on the charged particles within the mercury, causes them to move in a direction perpendicular to both the current and the magnetic field. As the mercury flows through the electromagnetic field, its charged particles experience this force, which can lead to complex motion patterns and even induce circulation within the liquid. Mercury's electrical resistivity plays a crucial role in determining the efficiency of the current flow and the overall effectiveness of the electromagnetic field in driving the motion. Lower resistivity leads to better current flow and a stronger interaction with the magnetic field, thereby enhancing the Lorentz force's impact on the mercury's motion. Additionally, the expansion coefficient of mercury comes into play as the liquid is heated due to resistive losses in the presence of the current. This expansion can cause the mercury to occupy a larger volume and potentially influence its flow behaviour within the system. In summary, the interplay between the Lorentz force, electrical resistivity, and mercury's expansion coefficient governs the motion of mercury in the presence of a DC current and an electromagnetic field, leading to intricate flow patterns and unique dynamical behaviour.

To outline the math for the motion of mercury in the presence of a DC current and an electromagnetic field, we can consider several fundamental equations and concepts, including the Lorentz force, Ohm's law, and the equation of continuity. These equations can be applied to describe the behaviour of the system under study.

1. Lorentz force: The Lorentz force F acting on a charged particle of charge q moving with a velocity v in an electric field E and a magnetic field B is given by:

$$F = q(E + v \times B)$$

For a current-carrying conductor, such as mercury, the current density J (current per unit area) can be expressed as:

$$J = nq\_v$$

where n is the number density of charge carriers, and q_v is their charge.

2. Ohm's law: The relationship between current density J, electric field E, and electrical conductivity σ (the inverse of resistivity ϱ) is given by Ohm's law:

$$J = \sigma E$$

3. Navier-Stokes equation: The motion of a fluid, such as mercury, can be described by the Navier-Stokes equation, which is a momentum balance equation:

$$\varrho(\partial v/\partial t + v\cdot\nabla v) = -\nabla P + \mu\nabla^2 v + F$$

where ϱ is the fluid density, v is the fluid velocity, t is time, P is the pressure, μ is the dynamic viscosity, and F is the external force acting on the fluid (in our case, the Lorentz force).

4. Equation of continuity: To account for the incompressibility of mercury and its expansion coefficient, we can use the equation of continuity:

$$\nabla\cdot v = 0$$

5. Thermal expansion: The volumetric expansion of mercury due to heating can be described using the coefficient of volumetric expansion β:

$$\Delta V/V = \beta\Delta T$$

where ΔV is the change in volume, V is the initial volume, and ΔT is the change in temperature.

To analyse the motion of mercury in the presence of a DC current and an electromagnetic field, these equations can be combined and solved numerically, taking into account the material properties of mercury (such as its electrical conductivity, dynamic viscosity, and coefficient of volumetric expansion). By solving the resulting system of equations, we can study the complex motion patterns and behaviour of mercury under the influence of electric and magnetic fields.

The proposed system utilizes the motion of mercury in the presence of a DC current and an electromagnetic field to create a self-sustaining electromagnetic field. The concept relies on the interplay between several physical phenomena, such as the Lorentz force, induced electromotive force (EMF), and the conservation of energy. Here's a detailed explanation of how this system can work:

1. Initiation: When the engine starts, an external DC current is applied, creating an initial magnetic field around the mercury. This current, combined with the motion of mercury induced by the electromagnetic field, sets the mercury in motion.

2. Mercury motion and Lorentz force: As the mercury moves through the magnetic field, it experiences a Lorentz force, which causes the charged particles within the mercury to be pushed perpendicular to both the electric field and the magnetic field. This force leads to an increase in the velocity of the mercury, and consequently, the rotation of the system.

3. Induced electromotive force (EMF): The motion of the mercury within the magnetic field induces an EMF in the system, according to Faraday's law of electromagnetic induction. This induced EMF generates a secondary current within the mercury, which can, in turn, contribute to the maintenance and enhancement of the magnetic field.

4. Energy conservation and feedback: As the engine reaches higher RPM, the kinetic energy of the rotating mercury increases. This increase in kinetic energy can be harnessed to maintain the DC current, effectively creating a feedback loop that

sustains the electromagnetic field with minimal energy input.

5. Short-circuit effect: The self-sustaining nature of the system can also lead to a "short-circuit" effect, where the energy generated by the induced EMF is fed back into the system, further enhancing the motion and the generated magnetic field. This effect can potentially result in a system that requires minimal external energy input once it reaches a high RPM.

However, it is essential to note that energy conservation laws still apply, and the system will likely have energy losses due to friction, heat generation, and other inefficiencies. These losses will eventually require additional energy input to maintain the system's operation. Nevertheless, if the system is designed efficiently, it may be possible to create a self-sustaining electromagnetic field that requires minimal external energy input once the engine has reached a high RPM.

In conclusion, the proposed system creates a self-sustaining electromagnetic field by utilizing the motion of mercury, the Lorentz force, induced EMF, and energy conservation principles. This system, if designed efficiently, could potentially operate with minimal energy input once it reaches higher RPM, making it an intriguing concept for further exploration and development.

Based on the concepts and principles discussed earlier, we can formulate a final hypothesis for the design of a system utilizing mercury in motion under the influence of a DC current and an electromagnetic field. The design aims to exploit the unique properties of mercury, such as its electrical conductivity, expansion coefficient, and density, to create a device with potential applications in various fields, such as energy conversion, propulsion, or magnetic levitation.

HYPOTHESIS

Hypothesis:

The proposed design, which combines the motion of mercury in the presence of a DC current and an electromagnetic field, can lead to the generation of a strong, stable magnetic field along with gyroscopic stability due to the circulation of mercury. This magnetic field, coupled with the centrifugal and centripetal forces experienced by the mercury, will allow for efficient energy conversion or levitation of objects placed above the device.

To test this hypothesis, a detailed computational model and experimental setup can be developed based on the mathematical framework outlined earlier. By simulating the system and measuring the generated magnetic field, the efficiency of the energy conversion, and the levitation capabilities, the hypothesis can be validated or refined.

Further research may explore different geometries, materials, and operating conditions to optimize the system's performance and identify the most suitable applications for the proposed design. The scalability, safety, and environmental impact of the design should also be considered as part of the investigation.

In conclusion, the proposed design utilizing mercury in motion under the influence of a DC current and an electromagnetic field presents an intriguing hypothesis for creating a device capable of generating a strong magnetic field and gyroscopic stability. By testing and refining the hypothesis through simulation and experimentation, it may be possible to develop a novel system with applications in various fields, from energy conversion to magnetic levitation.

RISK ASSESSMENT

Risk Assessment:

As we proceed with the investigation of the proposed design, it is essential to consider additional factors that may influence the system's performance and its practical applications. Some of these factors include the safety and environmental aspects, the challenges associated with handling mercury, and possible technological advancements that can be integrated to enhance the design further.

1. Safety and environmental concerns: Mercury is a toxic heavy metal that poses health risks and environmental hazards. Its handling, storage, and disposal must be managed carefully to minimize the risks associated with exposure and contamination. Any practical implementation of the proposed design should incorporate appropriate safety measures, such as protective enclosures and ventilation systems, to ensure that

mercury does not pose a threat to operators or the environment.

2. Challenges in handling mercury: Due to its high density and unique properties, handling mercury can be challenging. It may require specialized equipment and techniques to ensure proper containment, movement, and control within the system. Research into effective ways to handle mercury within the design, as well as innovative approaches to minimize the need for direct interaction, will be vital to the overall success of the project.

3. Technological advancements: As technology advances, it may be possible to integrate new materials, components, or methods into the proposed design to improve its performance, efficiency, and safety. For instance, the use of advanced materials with improved electrical conductivity or magnetic properties could enhance the effectiveness of the generated magnetic field. Similarly, incorporating more sophisticated control systems or sensors may enable better monitoring and control of the mercury flow, leading to optimized performance.

4. Alternative designs and applications: Exploring alternative design approaches or variations of the proposed system can lead to new insights and potential applications. For example, investigating different geometries, configurations, or the use of alternative fluids with unique properties may reveal new opportunities for innovation. Additionally, studying the proposed design's potential applications in various industries, such as

aerospace, transportation, or renewable energy, can help identify the most promising areas for future research and development.

In summary, as we continue to investigate the proposed design of a system utilizing mercury in motion under the influence of a DC current and an electromagnetic field, it is crucial to consider a wide range of factors, from safety and environmental concerns to technological advancements and alternative approaches. By addressing these challenges and exploring new opportunities, the proposed design can be refined and optimized, ultimately leading to a novel system with potential applications in various fields.

CONCEPT ELSEWHERES

The proposed system in this case involves using the kinetic energy of a flowing river to spin up a millstone horizontally like a spinning top. The spinning millstone, made of granite (a piezoelectric material), will be wrapped in copper wire to generate electricity using the piezoelectric effect. Here is a detailed explanation of how this system can work:

1. Harnessing river energy: To harness the kinetic energy of the river, a waterwheel can be installed in the river flow. The waterwheel is designed with blades or paddles that are driven by the river's motion, converting the water's kinetic energy into mechanical energy.
2. Gear system for torque and speed: The mechanical energy from the waterwheel is then transferred to a gear system. This gear system can be designed to increase torque and spin the millstone at the desired speed. The gear ratio can be adjusted

according to the required torque and speed for optimal energy generation.

3. Millstone and piezoelectric effect: The millstone, made of granite, will be spun horizontally like a spinning top. Due to the piezoelectric properties of granite, mechanical stress created by the spinning motion generates an electric charge within the material. This charge accumulates on the surface of the granite, creating an electric potential difference.

4. Copper wire and charge transfer: The spinning millstone will be wrapped in copper wire, which acts as an electrical conductor. As the millstone spins and generates electric charge, the copper wire collects and transfers the charge, creating an electrical current.

5. Charging and starting the system: To start the system, an initial energy input might be required to overcome inertia and get the millstone spinning. This could be achieved manually or using a small motor. Once the millstone reaches a sufficient speed, the river's energy should be enough to maintain the spinning motion and generate electricity.

6. Utilizing the generated electricity: The electricity generated through this system can be used to power electrical devices, charge batteries, or be fed into the electrical grid. Proper voltage regulation and control systems should be in place to ensure the safe and efficient distribution of the generated electricity.

It is essential to note that this system will be subject to energy losses due to friction, heat generation, and other inefficiencies. However, by optimizing the design and materials used, it is possible to create a sustainable energy generation system utilizing the kinetic energy of a river and the piezoelectric properties of granite.

CONCEPTS CONTINUED

Integrating this design into the water mains system can provide a novel approach to harnessing the kinetic energy of flowing water for electricity generation. Adapting the concept for use within the water mains requires addressing the unique challenges posed by the enclosed and pressurized environment. Here is a detailed explanation of how this system could be implemented in the water mains:

1. Turbine design: Instead of using a traditional waterwheel, an inline turbine designed for use in pressurized systems should be employed. The turbine's blades should be optimized for the water flow rate and pressure in the water mains to maximize efficiency. Axial flow, radial flow, or helical turbines could be considered, depending on the specific conditions of the water mains system.

2. Pressure reduction valve integration: In many water mains systems, pressure reduction valves (PRVs) are used to decrease the water pressure to suitable levels for distribution. These PRVs can be adapted to accommodate the turbine, allowing the turbine to harness the energy from the pressure drop.

3. Housing and sealing: The turbine and gear system should be housed within a sealed enclosure to prevent water leakage and contamination. The enclosure should be designed to withstand the pressure and water flow in the water mains system. Appropriate seals and gaskets must be used to ensure a watertight fit.

4. Granite millstone and piezoelectric effect: Similar to the previous concept, the spinning millstone, made of granite, will be connected to the turbine's output shaft. The mechanical energy from the spinning motion will create an electric charge within the granite due to its piezoelectric properties.

5. Copper wire and charge transfer: The spinning millstone will be wrapped in copper wire, which acts as an electrical conductor. As the millstone spins and generates electric charge, the copper wire collects and transfers the charge, creating an electrical current.

6. Power management and safety systems: The generated electricity must be managed through appropriate power management systems, including voltage regulation and control systems, to ensure the safe and efficient distribution of the generated electricity. Additionally, safety features such as pressure relief valves and surge protection should be implemented to protect the system from damage due to sudden pressure changes or electrical surges.

7. Maintenance and monitoring: A monitoring system should be in place to track the performance of the turbine, millstone, and electrical generation components. Regular maintenance will be required to ensure the system continues to operate efficiently and safely, including checking seals, bearings, and electrical connections.

Implementing this design in the water mains system has the potential to harness the energy from flowing water for electricity generation. However, it is essential to optimize the design for the specific conditions of the water mains system and to ensure the safety and efficiency of the system.

for more information or if you want to work with me, you
can contact me at
ashyoplayslol@gmail.com
or you can text
+44 07868 161 503

www.ingramcontent.com/pod-product-compliance
Lightning Source LLC
Chambersburg PA
CBHW070922220526
45467CB00004B/1506